T0318248

Commercializing
Nanotechnology

Commercializing
Nanotechnology

A Roadmap to Taking
Nanoproducts
from
Laboratory
to
Market

Hebab A. Quazi

CRC Press
Taylor & Francis Group
Boca Raton London New York

CRC Press is an imprint of the
Taylor & Francis Group, an **informa** business

First edition published 2021
by CRC Press
6000 Broken Sound Parkway NW, Suite 300, Boca Raton, FL 33487-2742

and by CRC Press
2 Park Square, Milton Park, Abingdon, Oxon, OX14 4RN

Visit the Taylor & Francis Web site at
http://www.taylorandfrancis.com

and the CRC Press Web site at
http://www.crcpress.com

ISBN: 978-1-138-03191-3 (hbk)
ISBN: 978-0-429-18891-6 (ebk)

Typeset in Times
by codeMantra

CONTENTS

PREFACE

During the past two decades, basic research activities in nanoscale science and engineering have increased exponentially, and some of them have great potentials for commercial applications. But, as of today, very few nanotechnologies-based commercial products are in the market.

The driving force in commercializing nanotechnology is innovation. Once a potential application is identified, then the challenge becomes creating a solution for delivering the nanoproduct to the ultimate end user. Commercializing nanotechnology is an end-to-end venture. It starts with research and development and continues through to prototyping, testing, manufacturing, and marketing. In addition to technical knowledge and knowhow, these activities require expertise in financing, marketing, and business management. Other critical expertise needed includes market entry and sustainable business operation.

Millions of dollars are being invested in basic research and development, but funding is almost nonexistent for taking these potentials to prototyping, testing, and commercialization. The university laboratories now engaged in nanotechnology research are normally not equipped to handle prototyping and testing that are essential for commercialization.

During the past decade, many nanotechnologies that have been developed in the research labs reportedly have excellent potentials for commercial use. But very few of these have been prototyped and built to test commercial viability. These steps are necessary before considering commercial ventures. The prototyping and manufacturing requires appropriate tools and equipment normally not available in the university laboratories.

Certain professional organizations such as the American Institute of Chemical Engineers (AIChE), American Society of Mechanical Engineers (ASME), and Institute of Electrical and Electronics Engineers (IEEE) are providing professional platforms for helping researchers publish their lab findings.

The product development and commercialization require available standards and guidelines; knowledge of technology readiness; innovative methodologies and techniques; product design, development, and testing; sustainability planning, risk assessment, and risk mitigation; and manufacturing, financing, and marketing. Commercialization also requires creating sustainable ventures including end-to-end solutions. Currently, there are nanotechnology research programs that have potentials for nanoproducts particularly in medicine, biomedical, defense, and energy.

Pathways to successful commercial ventures include identifying an application (product/market potential), defining the challenges in creating a solution, describing competitive advantages of the proposed solution, selecting a technology for application, assessing technology's current readiness level, developing commercialization plan, arranging funds or finances, building prototypes for demonstration, and manufacturing actual system for testing the viability of commercial sustainability.

Challenges to commercialization include few standardized tools and techniques; lack of awareness within professionals; few environmental, health, and safety regulations; very few nonexisting historical trends; inadequate reliable data for investors; limited risk mitigation guidelines; uncertain market trends; and few success stories.

This book covers current culture of commercialization. It also presents potentials for commercial products. The book also covers nontechnical areas such as strategic planning and implementation for commercial ventures. Because of fast potential growth in nanotechnology commercialization, we see the need for having this book updated every 2–3 years for sharing additional new techniques and methodologies with the readers.

ACKNOWLEDGMENTS

I am honored to dedicate this book to many who contributed to the development of nanoscience and technology and to those who helped create opportunities for commercial application of nanotechnology.

My heartfelt gratitude goes to my parents (Hemayetuddin and Amjuara) who raised me and helped me shape my personal life and gave the freedom to develop my independent thinking space.

I am indebted to my wife Selima for supporting me in all aspects of my life particularly in pursuing my professional activities for over 58 years and two daughters and their husbands (Shahina and Moinul) and (Tanima and Ayman) for helping me in nourishing my thoughts in writing this book. I am also indebted to my grandchildren (Samir, Riana, Sabir, Adam, Aden, and Ava) who made my life so colorful and rewarding.

I am indebted to my friends (Bob Kimmons, Bill Mitchem, and Nick Tillmann in particular) and my family for encouraging and supporting me in pursuing my passion in working with professional societies, particularly with AIChE, ASME, and ISA, and for sharing my acquired knowledge with the readers of this book.

Also, I will remain greatly indebted to those who will facilitate in helping commercializing nanotechnologies.

AUTHOR

Hebab A. Quazi has been providing technology and management leadership to US federal, state, and city government programs and to industries and utility companies within the United States and internationally. He also provided leadership on programs funded by the World Bank, the United Nations, the Asian Development Bank, and the US Departments of Defense and Homeland Security. He completed programs in Asia, Europe, and North and South America. For over 25 years, he was the President and CEO of a US technology company and vice president of operations for the US operation of a European technology company.

He has led multidiscipline teams for innovative solutions for challenging technological issues. His experience also covers research and development and technology testing and commercialization.

Dr. Quazi was a director at the AIChE Nanoscience and Engineering Forum and chaired several technical programs on nanomaterials manufacturing. He was also a technical director for the Nanotechnology Assessment and Analysis Program Support for the US Department of Defense. There, he tested nanomaterials-based corrosion and abrasion prevention paints at the US DoD Assets.

He was the chairman for the New and Developing Technology Subcommittee at the Fuels and Petrochemicals Division of AIChE. For several years, he was a consulting editor of AIChE's *Chemical Engineering Progress* (Monthly). He was appointed by the US Secretary of Commerce to the Environmental Technologies Trade Advisory Committee.

Dr. Quazi earned a PhD in applied science from England as a Commonwealth Scholar. He also completed a 3-year graduate business education at the University of California, Los Angeles. He is a registered professional engineer in the state of California. He has written over two dozen industry reports (customer-confidential) and articles that were published in *The Encyclopedia of Physical Science and Technology* and *Project Management: A Reference for Professionals.*

1

INTRODUCTION

During the past two decades, basic research activities in nanoscale science and engineering have increased exponentially, and some of them show great potentials for commercial applications. But, as of today, a very few nanotechnology-based commercial products are in the market. The driving force in commercializing nanotechnology is innovation. Once a potential application is identified, then the challenge begins with creating a solution, delivering it to the end users or customers.

Commercializing nanotechnology requires creating end-to-end solutions. It starts with research and development and then continues through prototyping, testing, manufacturing, and marketing. The efforts also require resources for financing, marketing, and business management expertise. Other critical steps include strategic planning for market entry and sustainable business operation.

Millions of dollars are being invested in basic research and technology development in university labs, but funding is almost nonexistent for moving these potentials to prototyping, testing, and commercialization. The university laboratories engaged in nanotechnology research are normally not equipped to handle prototyping and testing that are essential for commercialization.

During the past decade, many nanotechnologies that have been developed in research and development labs have excellent potential for commercial products. But very few of these have been prototyped and built to test commercial viability. These steps are necessary for any commercial ventures. The prototyping and manufacturing requires appropriate tools and equipment that are not always available in the university laboratories.

Few professional organizations such as the American Institute of Chemical Engineers (AIChE), American Society of Mechanical Engineers (ASME), and Institute of Electrical and Electronics Engineers (IEEE) are providing professional platforms for helping researchers presenting or publishing their lab findings.

The product development and commercialization require appropriate standards and guidelines; knowledge of technology readiness; innovative methodologies and techniques for product design, development, and testing; sustainability planning; risk assessment and risk mitigation; and financing, manufacturing, and marketing. Commercialization also requires creating sustainable values and ventures that include end-to-end solutions. Currently, there are a limited number of nanotechnology research programs that have potentials for commercial products in medicine, biomedical, defense, and energy.

We are expecting exponential growth in innovative nanoproducts. Roadmaps to successful commercial ventures include identifying an application (product/market potential), defining the challenges in creating a solution, describing competitive advantages of the proposed solution, selecting a technology for application, assessing technology's current readiness level, developing commercialization plan, arranging for funds (or finances), building prototypes for demonstration, and manufacturing few components or a system for testing the viability for commercial sustainability.

Currently, the barriers to commercializing nanotechnologies include few standardized tools and techniques; lack of awareness within financing professionals; few environmental, health, and safety regulations; almost nonexisting historical trends; inadequate reliable data for investors; limited risk mitigation guidelines; uncertain market trends; and very few success stories. There are only few industry standards and guidelines available for nanotechnology development and commercialization. These include ISO TC 229, ANSI-NSP ASTM E56, and IEC TC 113.

Chapter 1 includes highlighted features of each chapter of the book. Chapter 2 covers *Nanotechnology: Research and Development* activities. In recent years, the research and development activities in nanotechnology are progressing well. These activities will lead to certain commercial products in the future. Innovative techniques are being developed almost every day. We expect to see exponential growth in innovative application of nanotechnology commercial products.

From application point of view, we are seeing focus on performance improvement, such as more distance, more speed, and more capacity. Researchers get excited in innovative applications of nanotechnology such as clothing for solders, solar powered battery for communication, and drug delivery for cancer treatment. We are also observing that fast-growing cities demand nanotechnology-based devices for sustaining safe and healthy environment in the fast-growing population in the urban areas.

The current focus in nanotechnology research and development is primarily in basics involving "thermodynamics and nanoreactions" and "nanostructure and nanoparticles." During the past 3–4 years, approximately 50 PhD level research work were completed, and their results were presented at technical meetings or published in the professional journals. Most of the research activities were focused into the fundamentals to improve the understanding of the technology, particularly on thermodynamics and reaction characteristics. In the application areas such as in "bionanotechnology and human health," the research work was focused on general application areas such as "nanomaterials for biological applications" and "bionanotechnology for gene and drug delivery."

In the "energy and environment" area, the research programs were designed to potential applications in energy conversions and storage. Considerable research and development programs have been focused toward graphene and carbon nanotubes particularly looking into characterization, functionalization, and dispersion. Certain work is being conducted for electronics application as well. Also, quite a bit of work is being focused on "manufacturing and processing" techniques that will help stepping up into prototyping the nanotechnology before proceeding to manufacturing.

These research activities are being presented at appropriate conferences under appropriate topical themes. These themes include (1) thermodynamics and nanoreactions, (2) nanostructure and nanoparticles, (3) bionanotechnology and human health, (4) electronics, (5) energy and environment, and (6) manufacturing and processing.

Chapter 3 deals with *Commercializing Techniques*. The chapter starts with introducing the enabling techniques for commercialization. Then, technology readiness assessment methodologies are described. Prototype design and testing techniques and nanomaterials manufacturing processes are also introduced. The intellectual property protection techniques are also discussed.

Chapter 4 covers techniques for *Targeting Commercial Markets* for potential nanotechnology commercial products. Market sectors covered includes agriculture and food, bioengineering, defense, energy, engineered materials, environmental and safety, manufacturing, and medicine and healthcare.

Chapter 5 covers the *Commercializing Plans* that will be necessary for commercializing nanotechnology. The plans covered include (1) business plan, (2) sustainability plan, (3) regulatory compliance plan, (4) risk management plan, (5) financing plan, and (6) product launching plan.

Chapter 6 deals with *Creating Sustainable Values for Nanoproducts*. It covers the activities that are essential for developing sustainable commercial ventures. Action items such as entrepreneurial leadership to building commercial ventures, teaming for sustainability, and customercare essentials are covered in this chapter.

Appendix A1 covers Characterization of Graphene Nanomaterial and Microporous Materials; Appendix A2 provides Guidelines to Selecting Manufacturing Process(s) for a Nanoproduct.

Because of fast growth in nanotechnology commercialization, we expect to have this book updated and revised every 2–3 years to keep the readers current with new techniques and methodologies.

2

NANOTECHNOLOGY
Research and Development

2.1 INTRODUCTION

In recent years, the research and development (R&D) activities in nanotechnology are progressing well. These R&D work will lead to certain commercial products. Innovative techniques are being developed almost every day and deployed at the molecular level. We expect to see exponential growth in innovative approaches for taking the nanotechnology to commercial products.

We are seeing research activities on performance improvement of the devices that are desired by the fast-growing smart cities. Many fast-growing cities demand nanotechnology-based devices for sustaining safe and healthy environment in the fast-growing population in the urban areas.

2.2 CURRENT RESEARCH AND DEVELOPMENT FOCUS

The current focus in nanotechnology R&D is primarily in basics involving "thermodynamics and nanoreactions" and "nanostructure and nanoparticles." During the past 3–4 years, approximately 50 PhD level research work were conducted, and their results were presented at the professional conferences or published in the professional journals. Most of the research activities were focused into the fundamentals for improving the understanding of the technology, particularly on thermodynamics and reaction characteristics. In the application areas such as in "bionanotechnology and human health," the research work was focused on applications such as nanomaterials for biological applications and bionanotechnology for gene and drug delivery.

In the "energy and environment" area, the research programs were designed for potential applications in energy conversions and storage. Also, considerable R&D programs have been focused toward graphene and carbon nanotubes, particularly looking into characterization, functionalization, and dispersion. Also, certain work is being conducted for electronics as well. Considerable share of work is being focused on "manufacturing and processing" techniques, helping transitioning to prototyping the nanotechnology before proceeding to commercialization.

Some of these research results are presented at appropriate conferences under appropriate topical themes. Some of the topical themes have as many as six (6) presentations in each. Readers are encouraged to explore the details of these research presentations, where possible.

2.2.1 Nanoscale Particles, Phenomena, and Systems

Considerable work has been initiated in the nanoscale particles, phenomena, and systems particularly for investigating principles such as functions, characteristics, and thermodynamics. Between 2015 and 2019, over 70 research papers were presented at recent conferences. Some of the general headings of the technical sessions are presented in the following.

a. Chemical Engineering Principles for Nanotechnology,
b. Complex Fluids: Suspensions and Nanoparticle–Polymer Materials,
c. Functional Nanoparticles,
d. Microfluidic and Nanoscale Flows: Multiphase and Fields,
e. Nanoreaction Engineering,
f. Nanoscale Phenomena in Macromolecular Systems,
g. Nanoscale Structure in Polymers,
h. Nanostructured Polymer Films,
i. Nanostructured and Self-assembled Polymer Membranes, and
j. Thermodynamics at the Nanoscale.

2.2.2 Nanostructure and Nanoparticles

A large number of research work has been initiated in the nanostructure and nanoparticles including for investigating molecular structures for catalysis and polymers. Between 2015 and 2019, about

100 research papers were presented at recent conferences. Some of the general headings of these technical sessions are presented in the following:

a. Carbon-Based Nanostructured Membranes,
b. Catalysis with Novel Nanoparticles and Nanostructured Materials—Influence of Particle Sizes, Support Interactions, Control in Synthesis and Application,
c. Characterization of Engineered Particles and Nanostructured Particle Systems,
d. Functional Nanoparticles,
e. Graphene 2-D Materials: Synthesis, Functions, and Applications,
f. Graphene and Carbon Nanotubes: Characterization, Functionalization, and Dispersion,
g. Graphene and Carbon Nanotubes: Absorption, and Transport Processes
h. Graphene and Carbon Nanotubes: Separations, Materials and Applications,
i. Nanomaterials and Nanotechnology Sustainability,
j. Nanomaterials for Thermal-to-Electric Conversion,
k. Nanoscale Phenomena in Macromolecular Systems,
l. Nanostructured Particulate Systems,
m. Nanostructured Particles for Catalysis,
n. Nanostructured Thin Films,
o. Nanoparticle Coatings and Nanocoatings on Particles,
p. Nanoscale Science and Engineering in Biomolecular Catalysis,
q. Novel Nanoparticles and Nanostructured Materials for Catalysis – Influence of Particle Size, Support, Synthesis and Processing,
r. Phase Behavior, Rheology, and Processing of Nanoparticle Suspensions and Solutions,
s. Semiconducting Nanocrystals and Quantum Dots, and
t. Synthesis of Graphene and Carbon Nanotubes: Kinetics, Mechanism and Reactor Design.

2.2.3 Bionanotechnology and Human Health

A large number of research activities have been initiated in the bionanotechnology and human health including investigations in biomimetic and biohybrid materials and devices. Between 2015 and 2019, about 125 research papers were presented at recent conferences.

Some of the general headings of the technical topics are presented in the following:

a. Bionanotechnology,
b. Bionanotechnology and Micro-Scale Technologies,
c. Bionanotechnology for Gene and Drug Delivery,
d. Environmental Implications of Nanomaterials: Biological Interactions,
e. Magnetic Nanoparticles in Biotechnology and Medicine,
f. Mechanical Engineering at the Nanoscale is Aiding Cutting-Edge Applications in Oncological Diagnosis and Treatment,
g. Nanobiotechnology for Sensors and Imaging,
h. Nanomaterial Applications for Human Health and the Environment,
i. Nanomaterials for Biological Applications,
j. Nanotechnology in Medicine and Drug Delivery,
k. Nanotechnology and Nanobiotechnology for Sensors and Imaging,
l. Nanotechnology for Biotechnology and Pharmaceuticals,
m. Nanoparticles and Health,
n. Nanoscale Science and Engineering in Biomolecular Catalysis,
o. Nanostructured Biomimetic and Biohybrid Materials and Devices,
p. Nanostructured Scaffolds for Tissue Engineering,
q. Pharmaceuticals and Medical Applications – Novel Nanoparticles and Nanostructured Materials, and
r. Self-Assembled Biomaterials.

2.2.4 Electronics

A limited number of research work have been initiated in the electronics area including investigations in nanoelectronics, photonics, and photovoltaics. Between 2015 and 2019, over 30 research papers were presented at recent conferences. Some of the general headings of the technical topics are presented in the following.

a. Photovoltaics – Nanostructured Thin Film,
b. Nanoelectronic and Photonic Materials,
c. Nanomaterials for Photovoltaics,
d. Nanostructured/Thin Film Photovoltaics, and
e. Nanoscale Applications.

2.2.5 Energy and Environment

Considerable number of research work has been initiated in the energy and environment including investigations in novel nanoparticles, artificial photosynthesis, photocatalytic and photoelectrochemical reactions, energy conversion, and storage. Between 2015 and 2019, over 100 research papers were presented at recent conferences.

a. Energy and Environmental Applications – Novel Nanoparticles and Nanostructured Materials,
b. Environmental Applications of Nanotechnology and Nanomaterials,
c. Environmental Aspects, Applications, and Implications of Nanomaterials and Nanotechnology,
d. Fuels from Sun – Nanomaterials for Water Splitting, Artificial Photosynthesis, and Other Photocatalytic and Photoelectrochemical Reactions,
e. Nanoelectronic and Photonic Materials,
f. Nano-Energetic Materials,
g. Nanomaterials for Energy Storage,
h. Nanomaterials for Hydrogen Production and Fuel Cells,
i. Nanomaterials for Light Harvesting and Novel Photophysical Phenomena,
j. Nanomaterials for Thermal-to-Electric Conversion,
k. Nanostructured Particles for Energy Conversion and Storage, and
l. Nanostructured Thin Film Photovoltaics.

2.2.6 Manufacturing and Processing

Certain research work has been initiated in the manufacturing and processing including investigations in manufacturing processes and assembly. Between 2015 and 2019, approximately 50 research papers were presented at recent conferences.

a. Emerging Applications of Cellulose Nanofibrils (CNFs) in Composites,
b. Nanofabrication and Nanoscale Processing,
c. Nanomaterials Manufacturing,
d. Nanomaterials Synthesis and Self-Assembly Strategies,
e. Nanoparticle Coatings and Nanocoatings on Particles,

f. Nanoscale Structures in Polymers,
g. Nanostructured Biomimetic and Biohybrid Materials and Devices,
h. Nanostructured Polymer Films and Systems,
i. Nanowires: Synthesis, Processing and Applications,
j. New Composites with 2-Dimensional Nanomaterials,
k. Self and Directed Assembly at the Nanoscale, and
l. Templated Assembly of Inorganic Nanomaterials.

2.3 NANOTECHNOLOGY R&D PROGRAM PLANNING FOR COMMERCIALIZATION

Nanotechnology R&D programs for successful commercialization begin with detail planning, scheduling and controlling concepts. It involves development of enabling plans, application of appropriate techniques and tools, and adopting industry best practices. These enabling capabilities, techniques, and tools are covered in this book. Chapters 3, 5, and 6 provide the engineers, scientists, and managers with information and tools for advancing the appropriate nanotechnology from R&D phase of the program to commercialization.

2.4 NANOTECHNOLOGY COMMERCIALIZATION SUPPORT RESOURCES

Very few R&D laboratories are well equipped with all the instruments and equipment that are required for prototyping nanotechnology-based product development and for testing. Sometimes, in-house costs are prohibitive for these activities, and for that reason alone, the researchers and engineers have to look outside sources.

There are several facilities in the world that can be used for prototyping nanotechnology-based product before taking to commercial ventures. The readers are encouraged to open discussion with outside resources that are reliable and affordable. The product developers are encouraged to establish relationship with these resources for their continued needs, especially for materials characterization.

Materials characterization support: Materials characterization is essential for most of the nanotechnology commercialization program. New nanomaterials that have been developed during the past two decades brought greater opportunities in many areas of applications including in electronics, energy, and healthcare. Analytical services for nanoparticles including characterization, particle sizing, classification, image analysis, turbidity, and standardization are available from a number of sources.

Brookhaven National Laboratory, Center for Functional Nanomaterials (BNL-CFN): This is a US Government facility (http://www.bnl.gov/cfn) for conducting interdisciplinary nanoscience and technology development programs. This lab operates under the US Department of Energy. They work on collaborative programs with private sector.

University of Texas at Austin, Nano Science and Technology (UT Austin NST): This facility offers a wide variety of tools and expertise for R&D programs. They also work with private sector.

Alliance for NanoHealth (ANH) is a multidisciplinary, multi-institutional collaborative research center for medicine, biology, material science, computer technology and public policy. Member institutions include Baylor College of Medicine, The University of Texas M.D. Anderson Cancer Center, Rice University, University of Houston, University of Texas Health Science Center in Houston, Texas A&M University, University of Texas Medical Branch, and the Methodist Hospital Research Institute. The Alliance was formed to use nanotechnology for better diagnosis, treatment and prevention, and saving lives.

International Iberian Nanotechnology Laboratory: The International Iberian Nanotechnology Laboratory—INL, located in Braga, Portugal, was founded under an international legal framework to perform interdisciplinary research and to deploy and articulate nanotechnology for the benefit of society. The INL aims to become the worldwide hub for nanotechnology addressing society's grand challenges with specific emphasis on aging and well-being, mobility and urbanization, and a safe and secure society.

The work undertaken by the INL research center will have a significant impact on people's lives, as well as notably contributing to the development of our society at large. Areas of research in INL also include (1) semiconductors and nanoelectronics, (2) nanomaterials, and (3) nanomedicine.

3

COMMERCIALIZING
TECHNIQUES

3.1 INTRODUCTION

Commercializing concepts begin with applying appropriate techniques in developing the strategy, plans, and programs. First, it is important to assess whether the technology is ready for commercializing. So, the technology readiness assessment methodology and techniques are introduced. Once the technology is considered ready, then it can be moved from the research labs to prototyping and testing. Prototyping design requires technical readiness and financial strength. Most often, technology developer has to borrow money for this phase of the program. Private and commercial lenders require that the technology developer (borrower) is capable of providing colateral (security) and will be able to repay the borrowed capital. Alternatively, the owner can offer the lender certain ownership of the venture.

One other major step covered in this chapter is selecting nanomaterials manufacturing process to increase the viability of commercializing venture. Once the technological steps are established, then it is very important to protect the intellectual property rights of the venture through patenting. This chapter covers all these early steps to commercialization.

3.2 TECHNOLOGY READINESS ASSESSMENT

The Technology Readiness Assessment (TRL) guidelines are systematic approaches to assist entrepreneur and its team to establish the current status of the technology before proceeding to commercialization. The TRL assessment methodology is a tool that establishes the current maturity level of the technology and helps estimate the remaining work that should be carried out before the technology is considered commercially viable. These guidelines can be used to estimate the capital and timeline requirements for the commercialization program.

The TRL methodology was initially developed and used by the US National Aeronautics and Space Administration (NASA) for technology planning. Soon thereafter, other US government agencies and industries started using TRL assessment methodology. The US Department of Energy (DoE) developed TRL assessment methodology that is commonly used by the industry. The TRL level descriptions are stated in the following for guidance.

TRL 1 Basic Principles Observed and Reported
The research work begins with translation into applied research and development (R&D).

TRL 2 Technology Conceptualized, Developed and Application Identified
Basic concept is developed, and practical application is identified. Initial target application is speculative, without basic proof-of-concept or verification.

TRL 3 Experimental Work Done and Proof-of-Concept Established
R&D work are initiated. Analytical analysis is conducted, and laboratory experimentation is completed to validate analytical predictions.

TRL 4 Component Validation in Laboratory Environment
System (or technology) components are integrated to establish that these work together in a laboratory environment.

TRL 5 Component Validation in Relevant Environment
Basic components are integrated with reasonably realistic supporting elements so that they can be tested in a relevant environment.

TRL 6 System Prototype Demonstration in a Relevant Environment
System (or technology) prototype is designed, built, and tested in a relevant environment.

TRL 7 System Prototype Demonstration in an Operational Environment

Verification is established by demonstrating an actual system (or technology) prototype in an operational environment.

TRL 8 System Completed and Verified through Demonstration

System (or technology) has been proven to work in its final form and under desired conditions.

TRL 9 System Proven through Successful Operation

System (or technology) has successfully performed repeatedly in real-world environment.

3.3 PROTOTYPE DESIGN AND TESTING

Scientists, engineers, and managers tasked with technology commercialization are challenged with prototyping. In addition to addressing technical issues, prototyping requires expenditure of capital that must be raised from outside sources or needs approval from the company top management. The challenges include establishing convincing value proposition for needing the prototyping steps and then developing the financial justification in a business case for acceptance by the investor or by the company top management. The business case must include short-term and long-term benefits and risks assessment for mitigation or avoidance decision. The business case should include at least situational analysis and financial justification.

Situational analysis should provide clear statement on current status, proposed solution, alternatives, if any, and risks analysis with recommendation for risk avoidance or mitigation. Current situation analysis is the first step in finding the solution. It should be defined first in fairly details so the cost associated with each step can be estimated. This is essential so that capital expenditure and their justifications can be prepared for investors or management consideration and approval. The proposed solution also should be broken down in details so that it is easy for the approval authorities to understand the measurable benefits.

The capital–expenditure proposal identifies clearly the value-added benefits including net profits that can be achieved and the time that will take before the benefits can be realized. Alternatives considered, if any must be clearly stated along with their associated costs and potential benefits and risks. It is very important to demonstrate that fair and sound evaluation can be conducted.

All potential solutions, particularly requiring capital–expenditures have associated risks. The risks must be assessed clearly, and the degree of risks must be ranked. Risks analysis need to be carried out, where appropriate recommendations for risk aversion or mitigation are noted.

Financial analysis with justification is the core function of the prototyping business plan, and it must include return-on-investment (ROI) estimate on the capital requested for prototyping. The justification normally has two parts: (1) capital investment requirements including expenses and (2) financial benefits including ROI or payback period. The best case scenario would show large return with small investment. But in reality, the top strategy would be to select the one with most compelling benefits with good return that requires minimum investment. Consider the measurability of the proposed ROI as the first pick. To increase the probability of success in securing the approval of prototype funding, identify the hardware that can be reused in operation after prototyping.

Three options for prototyping are (1) restructuring existing prototyping methods, (2) increasing existing prototyping activities, and (3) rapid prototyping. Selection of appropriate prototyping option begins with a search for previous experience in outsourced or/and in-house prototyping. Collecting historical data for any options of prototyping for the past three years would assist, if available. This review provides a baseline for all potentials options.

3.4 SELECTING NANOMATERIALS MANUFACTURING PROCESS

Nanomaterial manufacturing processes are developed, tested, and validated for use in commercialization. Prudent evaluation should be carried out before selecting one for your commercialization venture. Some of the nanomaterials manufacturing processes used recently for specific applications are identified in the following.

3.4.1 Biological

High-dimension biological analysis of carbon nanotube toxicity

3.4.2 Combustion

Cellulose-assisted combustion synthesis of nanoparticles for catalytic application

3.4.3 Deionization

Colloidal assembly by capacitive deionization

3.4.4 Electrospray

Liquid–liquid electrospray: a high-throughput nanomanufacturing platform for synthesis of micellar nanocomposites

3.4.5 Flow-Synthesis

Continuous-flow synthesis of Ni-based nanocatalysts

3.4.6 Infrared Spectrometry

Molecular synthesis and characterization of nanoparticles using infrared spectrometry

3.4.7 Laser Deposition

Synthesis of p-type ZnO nanorods on an InP substrate by pulse laser deposition

3.4.8 Microwave

1. Parametric study on gold nanoparticles flow synthesis in a microwave-assisted reactor
2. Roll-to-roll production of nanomaterials using microwaves
3. Rapid microwave-assisted synthesis of zinc oxide nanoforest for solar cell application
4. Microwave-initiated nanomanufacturing toward energy applications

3.4.9 Morphology Manipulation

Novel techniques for production and morphology manipulation of MXene nanosheets

3.4.10 Nanofiltration

Characterization and performance evaluation of PVA nanofiltration membrane coupled with TiO_2 nanoparticles

3.4.11 Separation

Separation of double-decker-shaped silsesquioxanes condensed with multiple functional groups

3.4.12 Process Synthesis

Process optimization for the synthesis of gold and copper nanoparticles from mixed precursor solution

3.4.13 Pyrolysis

1. Scale-up of nanoparticle manufacturing by flame-spray pyrolysis
2. Numerical analysis of multicomponent catalysts production by double flame pyrolysis
3. Novel one-step liquid flame spray pyrolysis synthesis of N-doped TiO_2.

3.4.14 Vortex Mixer

Preparation of drug nanoparticles stabilized by polysaccharide surfactant using four-inlet vortex mixer

Selecting appropriate manufacturing process: The selection of an appropriate manufacturing process is not that simple. With experience, it becomes easier. The details of the roadmap are provided in Appendix A: "A Roadmap to Manufacturing Process Selection." Here are some guidelines for the first-time effort.

Step 1: **Considering the Options**
   Innovate
   Adopt
   Integrate
Step 2: **Review the Requirements**
   Industry standards and guidelines
   Process technology readiness

Past experience and trends
Testing requirements
Risk assessment and risk mitigation
Capital and operating costs.
Step 3: **Methodology and Roadmap**
Identify the application
Define needs
Select applicable technology/process
Assess technology/process readiness level
Describe manufacturing process system(s)
Establish competitive advantages for each system
Identify technology commercialization challenges
Assess risks (technical/marketing/financial)
Develop risk mitigation strategies
Evaluate licensing and permitting requirements
Identify resource requirements
Develop design and operating team
Establish capital and operating requirements
Step 4: **Establishing Competitive Advantages**
1. Optimized design and operation
   - Increased reliability and profitability
   - Secured sustained competitiveness
   - Competent management and operating team
2. Conservation of resources
   - Lowest manufacturing cost
   - Maximized wastes utilization
   - Minimized environmental pollution
3. Additional benefits (local/regional/global)
   - Platform creation for industrialization
   - Establish skills development centers
   - Support socioeconomic developments

3.5 INTELLECTUAL PROPERTY PROTECTION

Before completion of the R&D phase, the question must be addressed whether the results would be patentable or not. The consideration should be initiated and discussed within the organization. In addition to the scientists and engineers working on the R&D phase of the program, a team should be assembled from the manufacturing, marketing, management, and legal departments as a minimum to provide the guidance. A concept is new and is not enough to convince the examiner from the patent office. It is very important to understand the local patent office requirements. If possible, it would be good to understand the requirements for the regional and/or global patent rights.

The patent application should not only describe the invention but also differentiate it from similar concepts already patented. The patent rights when granted will also assign the territory and the duration it is valid for. The required fees for patent application can vary from country to country, region to region, as well as for global coverage.

4

TARGETING COMMERCIAL MARKETS

4.1 INTRODUCTION

Nanoinnovations have taken us looking at a large number of applications. We have learned from nature about certain nanostructured materials having special features, such as Gecko-foot having reuseable adhesive properties. Nanoscale is helping innovating unique and useful applications. We have started experiencing growth in nanoenabled products' innovation. Innovation is happening at the molecular level, and enormous applications of nanoparticles are under development. The applications cover many areas where innovative nanoproducts are poised to enter into the markets including in agriculture and food, bioengineering, defense, electronics, energy, engineered materials, environment and safety, manufacturing, and medicine and healthcare.

Commercialization involves innovation, adoption, and/or integration of emerging technologies such as nanotechnology. These steps require technology readiness assessment, prototyping and testing, manufacturing, sustainability planning, risks assessment and mitigation, market testing, financing, and customer acceptance. It also requires end-to-end solution. Commercialization of a nanotechnology product will have to go through at least the following steps:

1. Discovering a consumers' need or application
2. Defining the application clearly and estimating the market potentials
3. Creating the solution
4. Describing the competitive advantages of the solution offered
5. Selecting an appropriate technology or technologies
6. Assessing the current readiness level of the technology selected
7. Developing a sustainable commercialization feasibility plan

8. Arranging for funds or financing
9. Building a prototype for demonstration
10. Manufacturing few sample units for market testing
11. Testing the market and receiving customers' comments

The Commercialization Feasibility Plan must address at least the following key steps:

1. Evaluating the market (size, trend, and sustainability)
2. Identifying and evaluating competitions (competitive products, market shares, pricing trend, etc.)
3. Establishing the strategy for market entry
4. Identifying appropriate market entry methodologies
5. Matching own company strengths and weaknesses
6. Identifying market (local/regional/global as well as cultural) preferences, if any
7. Checking special situations such as economy of scales, logistic challenges, and regulatory climate, as applicable

Currently, barriers to commercialization include few standardized tools and techniques; lack of awareness within professionals; few environmental and health/safety regulations; nonexisting historical trends; inadequate reliable data for investors; limited risk mitigation guidelines; uncertain market trends; and few success stories.

On June 02, 2017, the Magazine for Small Science "Nano" reported on top five trendy nanotechnology applications. These are (1) more speed, (2) more distance, (3) more performance, (4) trendy clothes, and/or (5) eternal youth. Nanotechnology is definitely in the minds of many scientists and engineers but is trendy for everyone else creating imagination for all kinds of application.

Commercialization trends most often are created. Recognizing the potentials for taking nanotechnology from research and development stage to prototyping and testing is the challenge. Potential nanoproducts entering the commercial market will require end-to-end business solutions.

Viewing Broader Markets: The world of nanotechnology is vast. The market potentials beyond the current interests in the community are to be checked and identified. Certain applications of nanotechnologies may be overlooked initially, but these will be tagged once their sustainable benefits are recognized. We can visualize the nanoproducts that can impact our everyday life. These applications have huge potentials, but it needs considerable push to make these to commercially available products. Currently, some of these ideas are confined in academic research but can become commercial products in the near future. The opportunities are to be created, considered, evaluated, and/or developed under some of the following transformational domains that are normally overlooked.

Artificial intelligence: Artificial intelligence (AI) is being applied in the market place to secure greater benefits. AI in association with machine learning is now a powerful tool for the industries. AI, along with industrial Internet of things (IIoT), helped develop "Industry 4.0" platform. These innovative technologies are helping manufacturing plants becoming more efficient and competitive especially in the oil and gas, chemical, petrochemicals, mining, and manufacturing sectors.

Adoption of AI with nanotechnology will have wide applications. There are certain developing areas where AI converges with nanotechnology. Recent developments involve biology, optics, nanotechnology, and AI. In imaging area, AI brings benefit to scanning probe microscopy. Applications in this area have resolution issues particularly in imaging samples at nanoscale. AI can help making more efficient imaging system. From a simulation perspective, there are many different parameters that need to be correlated accurately to produce either an image or a moving depiction of a working system. AI can help better analyze the data, learn from the past, and produce a more accurate representation of the system. AI can also adopt nanocomputing. The computation is performed through nanoscale devices, and a wide range of AI methodologies can be applied to nanosensors.

Climatic challenge: Severe climatic changes are serious global challenges, such as a huge number of hurricanes, flooding, and heatwaves. These cause extreme casualties. Reversing the effects of severe climatic changes is very difficult but can be managed through reducing greenhouse gas emissions, primarily controlling carbon dioxide released to atmosphere. For over a century, fossil fuels are being used to generate power, and these produce a large quantity of greenhouse gases with carbon dioxide emissions causing global warming. During the past couple of decades, fossil fuels are being slowly replaced by renewable energy such as solar and wind energy. In this energy transition, nanotechnology can come to rescue.

While going through this energy transition, we are still faced with continued use of fossil fuels for many years to come. It makes sense to have technological solution for removing carbon dioxide gases entering the atmosphere. This can be done through carbon capture technology using nanosized membrane structures. Carbon capture is the process of removing carbon dioxide emissions from entering the atmosphere after fossil fuels are burned. Carbon capture technology relies on the principles that certain gaseous molecules are captured by an appropriate membrane. Membranes specially developed for the purpose can capture up to 90% of the carbon dioxide released in the combustion process. The membranes are reusable as the pores can be cleaned of any carbon deposits without damaging. This approach reduces greenhouse gas emissions to the atmosphere.

Digital transformation: Our world has been going through digital transformation for quite some time. By harnessing digital data, nanoproducts can undertake important roles in our social and economic domains. Nanoenabled data in real time can help manage smartly our health, energy, food, transport, infrastructure, water resources, education, security, communications, and finances. We can expect to see many nanotechnology products entering smart living domain in this digital age.

Innovative infrastructures: With expanding urban areas, construction is a vital part of its growth, particularly in its infrastructures. Nanomaterials can provide a positive impact in the composites used in the building materials. The concrete and cement required for these developments are already using certain nanomaterials. Nanomaterials have high tensile strengths and can be used in stress/strain gauges during and after the construction of buildings. Nanotechnology can also enable managing extreme heat, cold, and degradation.

Currently, certain road signs are storing data and helping create greater safety and security for modern cities. In addition, nanolatex inks can be used in road signs to track traffic movements improving the transportation security. Nanomaterials are used in ecopaints, making it anticorrosive coatings. Ecopaints can have the capability of removing pollutants from atmosphere. Nanomaterials have many other applications, including as an additive to asphalts for improving the road's resistance to wear.

Nanoenabled materials are already being used to make buildings more flexible and resistant to harsh environments against pollutants. Nanotechnology-enabled buildings will make it easier to be built with better resistance to the worst climatic perils. Nanoenabled materials are already being used to make buildings more flexible and resistant to atmospheric degradation. Future cities will continue to grow for accommodating increase in populations and sustainability. So, additional demands for nanomaterials will continue to increase.

Smart cities: Major cities around the world are growing fast. They are growing to accommodate ever-increasing populations, and many of their support systems are becoming highly automated. Nanotechnology is already impacting many aspects of city life, and their applications will continue to increase in the future years. Smart cities will enable complete technological transformation for enhancing their existing platforms. With nanomaterials (such as graphene), the industries are developing nanosensors and graphene batteries. This will help allowing optimum usage of time and energy.

Smart cities are already utilizing smart technologies in many of their systems including Internet of things (IoT), big data, and machine learning algorithms to enhance existing infrastructure performances. Nanotechnology is helping complete change on how we work, live, and play in the ever-growing urban environment. Smart cities are going to see the application of nanosensors providing extensive localized data points. Nanotechnology enables these tiny sensors to manage extraordinary amounts of data across multiple platforms. Some of the smart cities in the world are currently implementing 5G technology and are busy in laying underground superhighway of fiber-optic cables so it can transmit more data for their 5G networks.

Nanotechnology will enable smart cities to implement "Internet of nanothings" (IONT). It will provide support for remote environmental monitoring applications, personalized medicine, and early intervention in healthcare systems.

The revolution in healthcare will bring advanced point-of-care diagnostic tools at home allowing accurate self-diagnosis in real time using highly advanced nanoenabled mobile devices.

Conventional markets: Current thoughts on nanotechnology commercialization in the agriculture and food, bioengineering, electronics, energy, engineered materials, environment and safety, manufacturing, and medicine and healthcare are highlighted in the remaining sections of this chapter.

4.2 AGRICULTURE AND FOOD

Nanomaterials are slowly entering the agriculture and food markets. Food packaging industry has been trying to use nanomaterials for quite some time. Efforts are underway to use nanotechnology into foodstuffs, but safety standards must be met. Before any nanotechnologies are commercially materialized, the applications need to go through commercializing steps. The potentials are there for using nanoscale ingredients to improve the flavor, texture, and coloring of foodstuffs and to encapsulate vitamins, adding nutrients into beverages without altering the taste or texture of the drink.

Nanotechnology is being used to reduce the need for salt and sugar in our food. The application to foods also includes prevention of antibiotics entering in our food chain. We are likely to experience elimination of plastic particles in our food, animal, and fishes. This will ultimately create a food innovation ecosystem with positive impacts on our nature and environment. The initiatives that have been taken to develop nanoproducts in the agriculture and food area are listed here.

Bioplants: Research work is continuing to try to augment plants into bioplants with nanomaterials that would enhance their energy production capability and provide them with completely new functions, such as monitoring environmental pollutants.

Food freshness: Detection of freshness of foods stored in atmospheric temperature is very important to consumers for safe consumption. A tag has been developed having gel-like consistency that is inexpensive and safe that can be widely programmed to mimic almost all ambient temperature deterioration processes of foods.

4.3 BIOENGINEERING

A limited number of initiatives have been taken to develop nano-products in the bioengineering area. These potential applications are listed here.

Antimicrobial: A nanomaterial-based solution can be applied to treat cloths, carpets, and shoes that can kill a wide range of pathogens.

Biofilms: Engineers have coaxed bacterial cells to produce bio-films that can incorporate nonliving materials such as gold nanoparticles and quantum dots.

Early detection: Nanomaterials-based detection tools help early recognition and treatment of diseases.

Wearable monitoring: Researchers have developed a new stretchable antenna that can be incorporated into wearable health-monitoring devices to transmit data for patient monitoring or diagnosis.

4.4 DEFENSE

A limited number of initiatives have been taken to develop nano-products for application in this defined area. Some of these potentials are listed here:

Advanced camouflage: Researchers have progressed in testing nanobiomimicry applications in advanced camouflage and intelligent uniform.

Explosive detection: Certain nanotechnology can soon replace bomb-sniffing dogs. Researchers have identified a way to increase the sensitivity of a light-based plasmon sensor to detect minute concentration of explosives.

Performance enhancement: Thin-film nanotechnology is being tested for soldier's performance enhancement and safety.

Soldier protection: A nanomaterial-integrated textile is being tested for enhancing survival capability of soldiers in extreme environments.

Warmer clothing: Fabrics embedded with nanowires and hydro-gels can help soldiers stay warm and comfortable in colder climate.

Weight reduction: Printed nanoscale structures are being considered for weight reductions in defense systems.

4.5 ELECTRONICS

Several initiatives have been taken to develop nanoproducts in the electronics area. Some of these potential commercial applications are listed here:

Biochips: Nanoscale 3D printing technique is being used for micropyramids to build better biochips.

Broadband light detector: Researchers have found a way to control on how the material conducts electricity by using extremely short light pulses, enabling its use as a broadband light detector.

Brownian motor: Researchers have observed that "rocking" Brownian motor pushes nanoparticles around.

Data storage: Nanomaterial "graphene" pushes the possibility closer to data storage option.

Edible electronics: Electronics that dissolve or even be edible could be the answer to the ever-growing problem of e-wastes.

Electronic structure: Researchers have used a beam of ultraviolet light to look deep into the electronic structure of a material made of alternate layer of graphene and calcium.

Electronic devices: Flexible electronic devices such as e-readers could be folded to fit into a pocket. The approach involves designing circuits based on carbon nanotubes (CNTs) in place of rigid silicon chips.

Magnetic circuits: Nanoscale magnetic circuits are found to expand into three dimensions.

Nanoribbons: Graphene nanoribbons can be applied in molecular levels.

Semiconductors: Graphene combined with amorphous carbon increases its signal transmission efficiency and stability for use in semiconductor devices.

Sensors: Nanoparticles enable molecular electronic devices to work as sensors.

Transistors: Three potentials for nanomaterial use in electronics are considered in the following:

a. **Bilayer graphene**: Tests have been conducted successfully on the behavior of bilayer graphene to verify whether it could replace silicon transistors in electric circuits.

b. **Engineered graphene**: Research work indicates that the engineered bandgap brings graphene near to displacing silicon.

c. **Nanoring**: Nanoring transistor has high promise for future applications in electronics.

4.6 ENERGY

Energy market has been investigated reasonably well in finding appropriate applications of nanotechnology products. The opportunities include the following:

Batteries: The application of nanomaterials is starting to gain traction in the field of batteries. Detail investigations are yet to be carried out to assess the long-term safety, efficiency and charge/discharge cycle rates, but the potentials look very promising for many applications. The lithium battery with silver vanadium diphosphate electrodes has high potential. Sodium's abundance and low cost gives it an advantage over lithium for energy storage. Researchers have found a way to make sodium ion battery practical. They have developed an electrode nanomaterial that shortens the diffusion distance of sodium ions, giving such batteries much better rate performance than their predecessors.

Graphenes in batteries are examples of potential commercial application of nanomaterials. However, it will take time for testing the requirements in meeting regulations at least for safety and health. Additionally, sheets of graphene, graphene balls, carbon nanoscrolls, silver nanowires, and various lithium-based thin films could be used as electrode matrix. Many of these electrodes developed are known to store a larger amount of charge and have more efficient cycling rates and greater overall efficiencies compared with current graphitic-based electrodes. Cost of these nanomaterials can be the challenge to commercial application.

Efficient solar devices: The efficiency of solar cells is reduced as some of the absorbed light energy is lost as heat. Researchers have been looking to design materials that can convert more of that energy into useful electricity. They have paired up polymers that recover some of that lost energy by producing two electrical charge carriers per unit of light instead of the usual one.

Some self-contained materials work efficiently when dissolved in liquids. This opens up the possibility of having a wide range of applications such as solar energy–producing materials like ink for printing. Devices based on this multiplication concept have the potential to break through the upper limit on the efficiency of single junction solar cells. Currently, the efficiency is around 34%. The challenges go beyond doubling the electrical output of the solar cell materials. More-efficient current-generating materials could be added on to existing solar cell materials and device structures.

Energy storage: Nanomaterials will have wide use as the energy storage mediums in everyday electronics. Utilization of nanomaterials is considered safe over a long time periods.

Energy storage (mobile): Mobile energy storage technology has started using special graphene material to boost significantly the energy density of electrochemical batteries.

Graphene touch screen: Graphene has an excellent potential for use in touch screens for phones, tablets, laptop, watches, and interactive whiteboards. Having flexible touch screens will help enhance its use for wearable electronics and wearable technology. Graphene's optical transparency and flexibility helps in its high-tech applications.

Methanol fuel cells: Nanoparticle research has succeeded in proposing a new method to enhance fuel cell efficiency with the simultaneous removal of toxic heavy metal ions. The direct methanol fuel cell (DFMC) has been a promising energy conversion device for electrical vehicles. The new hybrid fuel cell technology is expected to propel the deployment of DFMCs.

Nanocolumns: Researchers have developed a technique for changing the color of a material by manipulating the orientation of nanostructured columns in the material. They have shown that the color of the material can be changed by using a magnetic field to change the orientation of an array of nanocolumns. The color-changing material has four layers. A silicon substrate is coated with a polymer that has been embedded with iron oxide nanoparticles. The middle layer is an aqueous solution containing free-floating iron oxide nanoparticles. This solution is held in place by a transparent polymer cover.

When a vertical magnetic field is applied beneath the substrate, it pulls the floating nanoparticles into columns, aligned over the pedestals. By changing the orientation of the magnetic field, researchers can change the orientation of the nanoparticle columns. Changing the angle of the columns shifts the wavelength of light that is most strongly reflected by the material. This has potential application in the energy sector.

Nanoresonators: Nanoelectromechanical resonators are used in all sorts of modern technology. These are found in robotics, medical tools, and environmental sensors. The researchers have demonstrated that the additional energy for the crystal oscillations comes in the form of heat caused by the electrical power.

Nanosensors: Graphene is slowly entering into sportswear that is durable, thermoregulating, and light weight. Incorporation of graphene in the sportswear has been shown to produce a running shoe with a much higher durability, stretchiness, and grip. There is also a large focus to realizing nanoinspired sensors into clothing. Flexible sensors are already into clothing, but they require built-in electrical circuits and a way of harvesting power from the body to keep them running and monitoring.

Nanowires for colder climates: Fabrics embedded with nanowires and hydrogels could help soldiers to keep warm and comfortable in colder climates. Fundamental research is underway to see if we can modify gloves for extreme cold weather. Scientists are developing smart fabrics that heat up when powered and can capture sweat. A study embedded a network of very fine silver nanowires in cotton, and it was able to heat the fabric by applying power to the wires. Soldiers would be able to dial the voltage up or down to vary the amount of heat they need. The system would be designed such that the uniforms could be lighter and thinner. The researchers are also incorporating a layer of hydrogel particles made of polyethylene glycol that will absorb sweat and stop the other layers of the fabric from getting wet.

Overheating prevention: The researchers describe a thermal transistor—a nanoscale switch that can conduct heat away from electronic components and insulate them against its damaging effects. Developing a practical thermal transistor could be a game changer in how we design electronics.

Silicon solar cells: Reducing the amount of sunlight that bounces off the surface of solar cells helps maximize the conversion of the sun's rays to electricity. So, manufacturers use coatings to cut down on reflections. Etching a nanoscale texture onto the silicon material itself creates an antireflective surface that works. The coatings in solar cells are for capturing fully every color of the light spectrum. Each color of light couples best with different antireflection coating, and each coating is optimized for light coming from a particular direction.

Solar cells efficiency improvement: Researchers have developed a comprehensive model explaining how electrons flow inside new type of solar cell. The model allows better understanding of such cells and helps in increasing their efficiency. In solar cells, some of the light energy is lost as heat.

Solar energy: Solar cells rely on semiconducting junctions to convert the solar energy into electricity. Many nanomaterials have been widely established as materials that can be used in these junctions. Currently, there are different types of solar cells that employ nanomaterials because they provide much greater conversion efficiency over traditional solar cells. Traditional inorganic solar cells composed of silicon and indium tin oxide (ITO) are slowly being replaced by nanomaterials such as graphene, quantum dots, perovskite nanomaterials, and 1D nanowire.

Triboelectric nanogenerator: Dragging feet across the carpet helps generate electricity. Sometime, there are even visible sparks. The phenomena could enhance electrical system in the house. One can even integrate this into a T-shirt, when walking and rubbing against the skin would generate energy to light up light-emitting diodes.

Sustainable energy grid: The insulation plastic used in high-voltage cables can withstand over 25% higher voltage when nanometer-sized carbon balls are added. This will assist by improving efficiency gain in the power grids for sustainable energy system.

Surface texturing: Nanostructured surface texture would prevent the reflection of light off silicon that would improve the conversion of sunlight to electricity. Etching a nanoscale texture onto the silicon materials would create an antireflective surface working as thin-film multilayer coating.

4.7 ENGINEERED MATERIALS

Engineered nanomaterial, graphene, has many targeted applications but still has many more potentials. Graphene is a two-dimensional material and is known for its versatility. Here are some of the targeted applications.

Graphene foam: Graphene foam combined with epoxy is substantially tougher than pure epoxy and is far more conductive than other epoxy composites. Epoxy filler brings better conductivity at the cost of weight and compressive strength, but the composite becomes harder to process. It replaces metal with three-dimensional foam made of nanoscale sheets of graphene. Easily interlocked between graphene and epoxy, it helps stabilize the structure of graphene.

Brittle smartphone screens: Scientists have developed a new way to make smartphone touch screens that are cheaper, less brittle, and more environmentally friendly. The new approach uses less energy and does not tarnish in the air. At present, ITO is used to make smartphone screens, and it is brittle and expensive. The primary constituent, indium, is also a rare metal and is ecologically damaging to extract. Silver is the best alternative to indium but is also expensive. Combining silver nanowires with graphene is still a two-dimensional material. The new hybrid material matches the performance of the existing technologies at a fraction of the cost.

Silver nanowires have been used in touch screens before, but it is the first time it got combined with graphene. Combining silver nanowires and graphene in a large scale is easier by spraying machines and patterned rollers. The addition of graphene to the silver nanowire network also increases its ability to conduct electricity.

CNT catalysts: Researchers used a new class of single-atom catalysts (SACs) supported on CNTs that exhibit outstanding electrochemical reduction of CO_2 to CO. Nickel–single-atom nitrogen-doped carbon nanotubes (NiSA-N-CNTs) are considered to have the highest metal loading for SACs. Single atom of nickel, cobalt, and iron were supported on nitrogen-doped CNTs for comparison.

Diamond-hard materials: Researchers have developed a material that is flexible and light weight as foil but is stiff and hard enough to stop a bullet on impact. Flexible and layered sheets of graphene temporarily become harder than diamond and impenetrable upon impact. By applying pressure at the nanoscale with an indenter to two layers of graphene, researchers transformed the honeycombed graphene into a diamond-like material at room temperature. Graphite and diamonds are both made entirely of carbon, but the atoms are arranged differently in each material, giving them distinct properties such as hardness, flexibility, and conductivity for electricity. The new technique allows manipulation of graphite so that it can have beneficial properties of a diamond under specific conditions.

Graphene desalinates: Removing nitrates from the water supply can normally be achieved through biological conversion, but it is a slow process. Use of palladium to catalyze the conversion of nitrate to nitrogen does speed up the process enormously. But this reaction suffers from the drawback as it produces harmful by-product ammonia. Researchers have assembled 2D materials with subnanometer slits that have potential for water desalination. The materials are made from graphene, hexagonal boron nitride (hBN), and molybdenum disulfide (MoS_2).

Graphene for electronics: Scientists used doped graphene that allows addition or subtraction of electrons from it by chemical means. The experiment revealed that a doped graphene absorbs a single photon, and it excites several electrons proportionally to the degree of doping. The photon excites an electron that rapidly falls back down to its ground state of energy. As it falls, it excites two more electrons on average as a knock-on effect. A photovoltaic device using doped graphene could show significant improvements in device performances.

Graphene membranes: Graphene is hydrophobic as well as hydrophilic. It is stronger than steel, flexible, bendable, and 1 million times thinner than a human hair. Graphene oxide membranes were shown to be completely impermeable to all solvents except for water. Researchers showed that one can tailor the molecules so that it passes through these membranes by simply making them ultrathin. Newly developed ultrathin membranes are assembled in such a way that pinholes are formed during the assembly and are interconnected by graphene nanochannels that produces an atomic-scale sieve allowing the large flow of solvents through the membranes. This helps greatly the applications of graphene-based membranes for seawater desalination and organic solvent nanofiltration (OSN).

Graphene nanoribbons: New study shows elegant mathematical solution for understanding on how the flow of electrons changes when CNTs turn into zigzag nanoribbons.

Graphene photovoltaics: Graphene has gathered tremendous popularity due to its extraordinary strength, light weight, and cost-effectiveness. Studies have hinted that graphene can also be used as a photovoltaic material, turning light into electricity.

Graphene for stronger silk: Spider silk is one of the toughest polymer fibers known. Its properties are enhanced even further if the spiders are allowed to ingest graphene and CNTs. The enhanced fibers could find uses in specialty high-performance fabrics, such as parachutes, or in biodegradable applications such as in sutures or medical dressings. Spider silk is stronger than conventional silkworm silk and therefore has received much research attention in recent years. It is among the best spun polymer fibers in terms of tensile strength, ultimate strain, and toughness even when compared with synthetic fibers.

Nanoinnovation in sanitation: Nanotechnologies bring potentials with innovation for transforming conventional to advanced sanitation systems for more than 2 billion people in the world and currently do not have access to adequate sanitation. Nanoinnovation is going to bring greater sustainable benefits in the sanitation area globally.

Nanomaterial for sensors: There is a large focus on commercially applying nanoinspired sensors into clothing. This has potential applications in both health and lifestyle monitoring. Flexible sensors can be made into clothing, but they require built-in electrical circuits and a way of harvesting power from the body and for keeping them running and monitoring.

Pacman disks: Use of nanomaterial as in Pacman disks could be an excellent potential application. For understanding the mechanisms, the researchers have used micromagnetic simulations to investigate vortex core switching device driven by an in-plane nanosecond current pulse. The current density could be reduced by 75% compared with that of a circular disk of same diameter and thickness.

4.8 ENVIRONMENT AND SAFETY

CNTs have unique transport properties that can benefit certain modern industrial, environmental, and biomedical processes. The applications can include large-scale water treatment and water desalination system, kidney dialysis, sterile filtration, and pharmaceutical manufacturing. Taking inspiration from biology, researchers have pursued robust and scalable synthetic membranes that either incorporate or inherently emulate functional biological transport units. Recent studies demonstrated successful environmental management and safety applications.

Authentication: Unique patterns made from tiny, randomly scattered silver nanowires have been created in an attempt to authenticate goods and address the growing problem of counterfeiting.

Carbon supersprings: Carbon supersprings are created by deforming a large number of CNTs.

Construction materials: Graphene's excellent physical properties made a significant impact in the construction industry. These properties include electrical conductivity, tensile strength, and optical transparency. Graphene is used in composites and as an additive to cement for high strength, self-cleaning, and flexural strength. It also helps making the cement environmentally friendly.

Concrete and paints: Nanomaterials including graphenes have been in use in the construction industry. Use of nanomaterials such as nanotitania, nanosilica, nanozinc oxide, and nanosilver in certain special paints to enhance their anticorrosive properties. Carbon nanofibers are used for reinforcing concrete roads in the snowy environment.

Faster photodiodes: Silicon nanostructures bend light to make faster photodiodes.

Heat insulation: Researchers have started using nanoparticles in heat-insulating surface layers that protect aircraft engines from overheating, which increases the service life of the coating by 300%.

Heat sink: New CNT sheets are top heat sink in performance.

Magnetic field sensors: Scientists have shown on how magnetism can manipulate the way electricity flows through a single molecule. This is a key step that would enable the development of magnetic field sensors for hard drives with a tiny fraction of their present size.

Magnetic graphene: Researches have successfully induced magnetism in graphene while preserving graphene's electronic properties.

Nanodynamite: Powerful explosives have been developed with nanomaterials. Coated nanotubes provide burst of power to the smallest systems.

Robotics: Robotic systems have led to mass assembly of nanostructures.

Sensors: A skin-like stretchy and transparent film of single-walled CNTs acts as tiny sensors that can measure the force applied on it.

Smart solar windows: Silicon nanowires enable to have smart solar windows.

Smart watch: Gallium nitride–based nanowires will have triple efficiency for smart watch and other displays.

Efficient nanodevices: Researchers have demonstrated that clean nanotubes have better electrical properties. Semiconducting nanotubes become more resistant to current along their length. Measuring the conductivity of nanotubes has never been straightforward, and the research team has investigated the basic science underlying the variability. The contaminants, such as residual iron catalysts, are difficult to clean. Contaminants from carbon and water can affect the conductivity of nanotubes for application in nanoscale electronics.

4.9 MANUFACTURING

Breakthroughs in nanotechnology applications are contributing to innovative solutions in industrial sectors. Benefits come primarily from creating nanomaterials that exhibit certain properties, such as stronger, lighter, and better conduction of electricity. These can help create new manufacturing processes including for additive manufacturing. The most promise for innovative manufacturing technologies is based on new hybrid materials and the nanotechnology that allows generation of very, very small features in larger surface areas.

Microscale assembly: Currently, there is no mechanism to assemble components across multiple-size scales. Nanoscale features in nanomaterials can be very challenging for assembling components that is very difficult to hold in a device. Researchers are using modulated surface energy to control the adhesion of flexible tools to manipulate assembly. They are developing a set of tools with varying sizes of different soft contact tabs that can pick up very small devices and then reposition them. The traditional method is to place the devices in very high-end equipment where tiny robotic arms do the assembly, one by one.

Nanoscale assembly: Many protein filaments are well studied and are already finding use in regenerative medicine, molecular electronics, and diagnostics. However, the very process of assembly (protein fibrillogenesis) remains largely unrevealed. A better understanding of this process is anticipated to offer new applications in biomedicine, pathogen detection, and molecular therapy.

The formation of protein filaments is highly dynamic and occurs over time and length, and it requires fast measurements with nano-to-micrometer precision. Many methods can meet these criteria, but the caveat is to measure in water and in real time. The challenge is compounded by the need to have a homogeneous assembly characterized by uniform growth rates of uniformly sized filaments. The study provides a measurement foundation for studying different macromolecular assemblies in real time and holds promise for engineering customized nano-to-microscale structures in situ.

Patterning: Both miniaturization and high-volume processing are important in producing affordable complex circuits used in many devices. Generating very small features over a very large area is very challenging and is not cost-effective. Functionally, semiconductor materials that are produced are not scalable at this time.

Scalable nanomanufacturing: Large-scale photolithography for semiconductors can use high-density small patterns (on polymers) in microelectronics processing. Current photolithography methods are extremely expensive, but effective. Researchers are working on an optical process that avoids the expensive methods. The process works by partially using the material's response to light while controlling the light, interacting with the material. By combining these characteristics, one can create very sharp features that are very small and low-cost scalable system. Once developed, it will have a large impact in how electronic components are manufactured and how much they cost.

Trading nanomaterials: Currently, nanocommodity exchange enables buying and selling nanomaterials globally. The exchange is a self-regulating organization delivering an electronic trading platform. This exchange is for listing accredited, inspected, and validated nanomaterials and nanoenabled commodities for physical delivery. Over 5,000 variants of engineered nanomaterials are listed there. The nanomaterials with certification and toxicology reports assist trading industry with focused knowledge of nanoscience and nanotechnology.

4.10 MEDICINE AND HEALTHCARE

Nanotechnologies have made significant advances in the medicine and healthcare sector. Applications include diagnosis, drug delivery, treatment, imaging, artificial organ support, and nanomedicine. Healthcare is going through a revolution due to nanomedicine, and this will help create a sustainable healthcare system for urban environment.

There have been significant advances in the treatment of some of the most disabling diseases with nanotechnology. Nanomedicine is also changing healthcare systems with new clinical uses of drug delivery approaches, mobile diagnostics, new therapies, nanovaccines, antimicrobial treatments, implants, and prostheses. Telemedicine and telesurgery could also be brought about by advances in sensor networks and nanomaterial optics. Pandemics and plagues can be monitored and controlled using nanosensors, new point-of-care diagnostics, and nanomedicines. Certain innovative diagnostic tools and treatments have been approved for diseases such as Ebola and the Zika virus. Potential applications are covered in the following:

Antibiotics: Nanotechnology-assisted systems enable combating bacteria's defenses, allowing antibiotics to work.

Artificial retina: Special graphene properties (thickness, transparency, conductivity, and tensile strength) can produce key elements of an artificial retina.

Brain disease monitoring: Most damaging brain diseases can be traced to irregular blood delivery to the brain. Researchers have employed lasers and CNTs to capture unprecedented look at blood flow through a living brain.

Burn victim caring: Tests report the development of a novel, ultrathin coatings (nanosheets) that *can cling to body's most difficult-to-protect contours and keep bacteria away from* burn victims.

Controlled drug release: Nanotechnology and supramolecular chemistry helped develop complex nanodevices for biomedical applications making it possible to control drug release, signal transduction, and sensing. The device senses the presence of a specific molecular input from a known disease biomarker, a certain gene, enzyme, or hormone in blood or tissue. The research demonstrated the working principle on three antibodies, including HIV, and has used nucleic acids as model drugs. Advanced DNA chemistry can possibly help design slingshots that fire a variety of drug molecules. As the work advances toward human applications, the research will begin preclinical testing of specific drugs against specific diseases.

Destroying cancer cells: Gold nanoparticles are currently being used to treat cancer. Infrared waves heat up the gold nanoparticles that in turn attack and destroy everything from viruses to cancer cells. Researcher has developed a nanoparticle that would serve as a virus decoy and chemically attract viruses to attack it rather than the healthy cells.

Disease monitoring: Wearable vapor sensors are being developed for continuous disease monitoring for diabetes, high blood pressure, anemia, or lung ailment. The new sensors would be able to detect airborne chemicals either inhaled or released through skin. This would be the first wearable to pick up a broad range of chemicals, rather than physical attributes.

Drug delivery: The liquid graphene oxide crystal droplets open up possibility for their use in drug delivery, disease detection and control, and magnetic field controls for drug delivery.

Extraction of cancer cells: A new method has been developed to effectively extract and analyze cancer cells circulating in patients' blood. This is a postal stamp–sized chip (NanoVelcro) with nanowires 1,000 times thinner than human hair, coated with antibodies that recognize cancer cells.

Fighting influenza virus: Researchers observe several potential applications for the nanosheets, ranging from a nasal spray that can help people avoid their annual fight with the influenza virus. Also, the nanosheets can offer great sensing capabilities. These nanosheets can be scaled up from cheap building blocks. The benefits are twofold as the proteins do all the important catalytic and structural work in our bodies.

Fighting infections: A novel molecular slingshot has been developed that harnesses the human immune system to fight infection with lethal precision. The slingshots are simple and effective as these are designed to target and neutralize infectious pathogens with laser precision. The molecular slingshot uses a strand of synthetic DNA to which an infection-fighting drug compound has been affixed. Each end of this strand is also chemically programmed to recognize and bind together with specific infection-fighting antibodies to viral or bacterial threats. The DNA strand stretches across the gap between the two arms of the antibody, creating a rubber band–like firing mechanism. As it stretches, the band releases the drug compound at the exact site of infection.

Immunotherapy: Researchers have developed immunotherapy with nanotechnology that promotes organ transplant acceptance.

Injectable bandage: Researchers have developed synthetic polymers, known as peptoids, and a method that allows building the *polymer* with one monomer at a time. This allows programming chemical information into the material as these are developed. One advantage is that these peptoids can spontaneously self-assemble in water using fundamental attractive and repulsive forces found in physics. This breakthrough helps making a sequence-defined polymer superfast with superefficient yields.

Nanomesh antibiotics: The fight against global antibiotic resistance has taken a major step forward discovering a concept for fabricating nanomeshes as an effective drug delivery system for antibiotics. Health experts are increasingly concerned about the rise in medication-resistant bacteria. Researchers have produced a nanomesh that is capable of delivering drug treatments. Effectiveness of the nanomesh for two antibiotics (colistin and vancomycin) was added together with gold nanoparticles to the mesh, before they were tested. To deliver to a specific area, the antibiotics were embedded into the mesh produced using electrospinning. This technique gained considerable interest in the biomedical community as it offers promise in many applications including wound management, drug delivery, and antibiotic coatings.

Nanocrystal heals injury: Synthetic material for grafting may yield better outcomes than traditional methods. Researchers have created a material and used it successfully in preliminary animal trials. Three components create a synthetic bone-like material which is then drilled into the femur and tibia. Researchers have found a way to integrate an artificial ligament with native bone.

Nanoparticles destroy superbugs: Superbugs are evolving too rapidly to be counteracted by traditional drugs. It produces a chemical that makes bacteria more vulnerable to antibiotic (attacking superbugs). Traditional treatment steps have lost their edge against hard-to-kill microbes. Antibiotics may have a new teammate in the fight against drug-resistant infections. Researchers have engineered nanoparticles to produce chemicals that render bacteria more vulnerable to antibiotics. These nanoparticles help combat pathogens that have developed resistance to antibiotics.

Nanopore-released medication: Gelatin is used in the pharmaceutical industry to encapsulate active agents. It protects against oxidation and overly quick release. Tiny pores in the material have a significant influence on oxidation and quick release. Custom-tailored gelatin preparations are widely used in the pharmaceutical industry. Medications that do not taste good can be packed into gelatin capsules, making them easier to swallow. Gelatin also protects sensitive active agents from oxidation. Often the goal is to release the medication gradually. In these cases, slow-dissolving gelatin is used.

Nanosensors for detection: Researchers have developed exciting research paths in nanotechnology and supramolecular chemistry for complex nanodevices in biomedical applications such as controlled drug release, signal transduction, and sensing. These are aimed at cancer cells, viruses, and other diseases that have abnormal level of certain gene, enzyme, or hormone in blood or tissue.

The potential human impacts of these studies are significant. The ability to deliver drugs precisely where they are needed in the body enables doctors to use smaller, less toxic doses of stronger drugs. This way, drugs can take out the disease without taking out harming the patient. Lowering the physical, emotional, and financial impacts of treating life-threatening diseases such as AIDS and cancer also reduces the burdens of these widespread diseases on patients, their families, and society as a whole.

Nanotechnology-guided cancer surgery: Researchers have developed a new way to selectively insert compounds into cancer cells—a system that will help surgeons identify malignant tissues and then, in combination with phototherapy, kill any remaining cancer cells after a tumor is removed. If few malignant cells remain, they will soon die. By adjusting the intensity of the light, the action of the compound can be controlled and optimized to kill just the tumor and cancer cells. This research was carried out with ovarian cancer cells.

Nanopathologies: Variation in cancer risk among tissues is attributed to environmental factors or inherited predisposition. Nanopathologies can help understanding the medical observations. It is very important to understand that the pathogenicity of incidental nanoparticles is very similar to that of engineered ones.

Nanoparticles in the clinic: Nanoparticle drug delivery systems have been used in the clinic since the early 1990s. Since that time, the field of nanomedicine has evolved alongside growing technological needs to improve the delivery of therapeutics. Over these past decades, newer generations of nanoparticles have emerged that are capable of performing additional delivery functions, which can enable treatment via new therapeutic modalities. Many new generation nanoparticles have reached clinical trials and have been approved for various applications.

Delivery of the synergistic combination of daunorubicin and cytarabine is enabled by the nanoparticle platform since the encapsulated ratio of drugs is able to interact with target cells upon release. Each drug exhibits distinct pharmacokinetic profiles, and they are metabolized at different rates.

Over 45 different nonapproved nanoparticles (liposomes, polymeric, micelles, albumin bound nanoparticles, and inorganic nanoparticles) were listed as active in a total of over 80 different clinical trials (mostly for the treatment of various cancers but also for radiation exposure, arthritis, pneumonia, amyloidosis, hepatitis, and fibrosis). Of these 80 trials, 28 have since been completed with 12 being terminated early. Of the 45 different nanoparticles, 7 possessed targeting functionality, and 6 offered stimuli responsive functions. Since 2016, search revealed that 18 new nanoparticles have entered clinical trials. Of these 18 nanoparticles, 12 are liposomes and 17 are indicated for cancer (15 being for treatment and 2 for imaging).

In summary, nanoparticle drug delivery systems offer many advantages over their free drug counterparts and can fundamentally change how therapeutics are delivered and enable the development of novel treatment modalities. Nanoparticles also face unique challenges related to their biological, technological, and clinical limitations that must be addressed to achieve consistent clinical impact.

Prostate cancer treatment: Researchers have developed a way to treat prostate cancer by restoring tumor suppressors, based on preclinical models in the lab. The approach represents the convergence of nanotechnology and biology observed in about half of metastatic castration-resistant prostate cancers.

Targeting pathogens: Researchers have created a two-dimensional sugar-coated nanosheet that mimics the surface of cells and can selectively target pathogens such as viruses and bacteria.

Tumor cells extraction: Researchers have developed a new method for effectively extracting and analyzing cancer cells circulating in patients' blood. Circulating tumor cells are cancer cells that break away from tumors and travel in the blood, looking for places in the body to start growing new tumors called metastases. Capturing these rare cells would allow doctors to detect and analyze the cancers.

Virus detection: Nanomaterials assist identifying Zika virus in minutes. Current protocol requires blood samples to be taken to a testing lab and refrigerated before testing. Advanced biosystem with gold nanorods on paper detects the virus in minutes.

5

COMMERCIALIZING PLANS

5.1 INTRODUCTION

Commercializing nanotechnology from laboratory to market involves innovation, adoption, and sometime integration in an existing business process. According to Steve Jobs "Innovation has nothing to do with how many R&D dollars you have... It's not about money. It's about the people you have, how you're led, and how much you get it." Dale Carnegie's guide to innovation includes finding ideas, finding solutions, finding acceptance, implementing, following up, and evaluation.

Commercializing involves value transformation to commercial ventures creating end-to-end solutions. Innovation, adoption, and integration require certain intermediate steps including assessing technology readiness, prototyping and testing, risk assessment and mitigation, sustainability planning, manufacturing, financing, marketing, and customer acceptance.

The nanotechnology development, adoption, and insertion in a business process or venture need to go through a stepwise roadmap including the following:

1. Identifying and/or defining the issue or problem
2. Developing or creating a solution
3. Describing competitive advantages of the proposed solution
4. Selecting a technology or technologies for the proposed solution
5. Assessing its current technology readiness level
6. Developing commercialization plan
7. Arranging appropriate financing
8. Building prototypes for testing
9. Manufacturing or building hardware or system
10. Testing the market and collecting customer feedback

The critical factors for technology commercialization roadmap are prototyping, testing, and identifying the challenges in technology insertion in an existing venture. On the other hand, creating values for a complete venture depends on financing, marketing, securing resources, and development of management capability.

Nanotechnology is an emerging technology, and its commercialization along with other appropriate emerging technologies can sometimes bring mutual benefits and be advantageous. When considering integration or insertion of nanotechnology into an existing venture, considerations must be focused on additional benefits that can be had from emerging technologies such as digitization, machine learning, digital twins, smart sensors, artificial intelligence, advanced analytics, industrial Internet-of-things, and cybersecurity.

Also, for nanotechnology product insertion in an existing manufacturing plant, considerations must be given on industry priorities. These priorities include improving reliability, operability, competitiveness, and profitability. The industries also look into eliminating interruption and wastes. Where applicable, the manufacturing plant should select agile methodologies and techniques.

While designing a product, the industry needs to maximize its benefits from digitizing real-time data and optimizing for sustainability. The industry target should be attaining top operational performance and maintaining market leadership. When considering technology platform, industries would select smart manufacturing technologies. They also should prefer to be with a platform that allows them making continuous improvement of their operation, where possible.

5.2 BUSINESS PLAN

Considerable numbers of books have been published on developing business plans. These books describe on how to develop a complete business plan for a venture. A business plan can involve details and require considerable time and financial resources to prepare. At the early stage of commercialization of nanotechnology venture, it is very important to start with a brief, but precise "summary" of ideas for guidance, reference, and sharing with others. Such summary business plan should include at least problem identification and definition, solution (product/services) proposed, target market(s) with customers, sales/marketing strategy, business model, identified competitions, competitive advantages, financial projection, resources (technical, management, marketing, financial, regulatory), and target schedule. It will be good to have a paragraph summarizing the business plan. The summary business plan needs to include a section on how you plan to raise capital for supporting the initial years of your commercializing venture.

The planning section would tell us what needs to be done and when and what benefits it brings for satisfying overall objectives. The planning document is sometimes called project execution plan (PEP). This planning document normally consists of the following steps:

a. Definition of the tasks to be performed
b. Statements on the objectives of the venture
c. Identification of potential problems
d. Lists of assumptions
e. Defining the objectives
f. Statements on strategies
g. Identifying sequence of the tasks
h. Establishing required resources
i. Reviewing and finalizing the plan
j. Establishing probability of success
k. Finalizing the plan, after detail review
l. Reviewing the plan often and improving, as background data do change

It is prudent to prepare an early work schedule. The schedules establish the activities to be performed in sequence. The scoping and tasking prioritization include the following:

a. Task identification
b. Sequencing
c. Logical sequence of tasks
d. Allocation of resources
e. Duration of each task
f. Identification of critical tasks/critical path
g. Levelling of resources
h. Preparation of bar charts

It will be a prudent approach to review the plan at least every 6 months and adjust the plan where appropriate. In reality, situations that influence and control commercial ventures do change frequently.

5.3 SUSTAINABILITY PLAN

The success of a commercializing venture based on an emerging technology is to be monitored, adjusted, secured, and assured often. Successful nanotechnology commercialization venture must be sustainable. Critical components in commercialization roadmap include as a minimum technology development, prototype design and testing, financing, manufacturing, regulatory compliance, and customer acceptance. The management team must have adequate knowledge and experience to adjust the action plan as the situation changes. All of these issues must be a part of sustainability plan.

The sustainability plan should address at least the following issues:

1. Technology development
2. Regulatory compliance
3. Testing and manufacturing resources
4. Financing (as/when needed)
5. Market intelligence
6. Management continuity
7. Third-party resources for hire
8. Economic conditions

The sustainability plan could include a "commercializing project life cycle roadmap" identifying the pathways to follow for commercializing the nanotechnology. The life cycle roadmap should include the following:

a. Technology/business strategy
b. Feasibility requirements
c. Revenue optimization criteria
d. External resources for validation
e. Permitting guidelines
f. Financing requirements and appropriate sources
g. Constructability analysis
h. Product launching techniques
i. Customer feedback

5.4 REGULATORY COMPLIANCE PLAN

Currently, there are few industry standards and guidelines available for nanotechnology development and commercialization. These include the following:

a. ISO TC 229 Nanotechnologies
b. ANSI-NSP (Nanotechnology Standards Database)
c. ASTM E56 Committee for Standards
d. IEC TC 113 (Nanotechnology Standardization for Electrical & Electronic Products & Systems)
e. TAPPI Nanotechnology
f. BAM NanoScale Reference Materials Database

5.5 RISK MANAGEMENT PLAN

Commercializing emerging technologies encounter considerable risks. These risks are to be recognized diligently, and their effectiveness is to be identified, studied, and classified (low, medium, and high impacts). Once identified, analyzed, and classified, efforts are to be made to mitigate the risks. Mitigation of risks involves aversion, minimization, and developments of alternatives. Risk analysis and mitigating processes are well recognized, practiced, and published. The most challenges are the risks that are yet to be identified. Efforts should be made in reviewing the risks frequently, and steps should be taken to avert or minimize risks, where possible.

Process risk management (PRM) guidelines are a part of industry's ongoing commitment primarily in the areas of environment, health, and safety (EHS). Most of the countries in the world have EHS regulations and guidelines applicable for their industries. Investors in emerging technology ventures are careful and sometimes have their own experts in performing their own analysis.

The industry guidelines provide the basis for consistent management of manufacturing facility's operational hazards and risks to the plant personnel. Sometimes, each manufacturing company develops its own guidelines and practices. These guidelines are practiced in all of the facilities for consistency. The manufacturing industry adjusts its guidelines frequently after appropriate reviews of its practices. The program review and implementation schedule is set by the industry.

The risk management system (RMS) can be different for each manufacturing facility based on operational processes used in each. In the United States, manufacturing facilities are required to have injury and illness prevention plan (IIPP). Each facility is required to provide RMS and IIPP training to the employees. Usually, the manufacturing industries use outside resources for risk management and preparation of RMS and IIPP.

5.6 FINANCING PLAN

During the early stage of technology development and testing, the required funds come from the owner. Also, owner's family and friends are the sources of initial financial support. Once the technology has been tested and reports are considered bankable, then borrowing of funds can have little flexibility. Equity crowdfunding is an option, an online offering of new venture's securities to a group of individuals for investment. This option is guided by country's regulations on financial and securities. The other sources of financing are the conventional private funding individuals. These individuals will ask for significant portion of the assets or shares. Once the prototyping and market testing are done, then the corporate or institutional investors would be interested.

To attract investors, it is very important to prepare a believable sales/revenue forecast. Quite often, the owner of the commercializing venture is not equipped to develop the successful financial plan. It may be prudent to retain a financial advisor to work with the owner. The financial advisors are knowledgeable about investor's preferences in all segments of private and public capital markets.

The feasibility of financing will be based on answering certain questions. These are as follows:

1. What are each shareholder's reasons for participating in raising capital?
2. What share of ownership does each existing shareholders wish to retain?
3. What level of risk each existing shareholder will accept?
4. Are the cash flow forecasted is sufficient to justify each shareholder's investment?
5. Will the anticipated cash flow be sufficient to attract capital from outside?

The tax structure(s) in the appropriate country will have significant impact on attractiveness for the investment.

5.7 PRODUCT LAUNCHING PLAN

The product launching plan (PLP) is the last and final plan for the commercialization venture. If the PLP is not developed properly, the commercialization venture most likely will not be successful, even after successful implementation of all other plans. The PLP should cover at least the following steps:

1. **Objective**: The purpose of the PLP phase is to introduce the manufactured product to the potential customers and collect their feedback on presentability, packaging, pricing, usability, satisfaction, desirability as repeat buyer, and suggestions for improvement, if any.
2. **Planning**: Much consideration, time, and resources are to be devoted for planning the PLP.
3. **Schedule (phasing)**: The PLP efforts should be scheduled and if necessary should be phased for convenience.
4. **Control criteria**: Identify the specific data that must be collected or gathered for control. Characteristics of control issues must be defined and planned from the beginning.
5. **Staffing**: Identify the key staff, and their activities must be identified and allocated.
6. **Documentation**: Develop the forms including data collection formats.
7. **Customer feedback**: The customer feedback should be recorded in predesigned forms.
8. **Budget**: Allocation of financial resources should be set before the PLP phase.
9. **Reporting**: Identify the key officials that should review the data collected and provide their comments for next action(s).
10. **Authority**: It is very important to identify the officials who have the authority to review and approve the PLP. Also, officials are to be designated for authorizing any changes.

6

CREATING SUSTAINABLE ·VALUES FOR NANOPRODUCTS

6.1 INTRODUCTION

Commercialization involves innovation, adoption, and/or integration of nanotechnology in a business venture by creating values and competitive advantages for sustainable operation. These business ventures need to have effective end-to-end solutions for sustainability and successful commercial operation. Last five chapters dealt with creating values and developing competitive advantages for sustaining commercial venture in a competitive environment. In technically dominated market sectors, resources for nontechnical aspects are usually forgotten or marginalized. This chapter includes discussions on certain essential components of commercialization that are mostly nontechnical but are essential for any successful commercialization.

In commercializing a nanoproduct for a specific market sector, certain nontechnical skills and expertise are essential. It is often observed that the key essential expertise such as marketing in a narrow market segment, highly technically oriented product, or launching a product inappropriately can miss the target customers completely. Also, inability to understand the customer requirements can also be a cause of business venture failures. Depending on the market segment, a marketing specialist with appropriate experience should be hired. The experts are not available all the time, especially when you need them.

For creating or supporting a successful venture involving nanotechnology commercialization, adequate planning with details is essential for the total program. Lot of time, it is easy to miss the required tasks in a critical stage of the venture. This detail plan is very important to assess project status at any time, specifically for understanding when to change the course of actions, if warranted.

The detail plan is required for each phase so that one could have the total control of the program.

For improving the success factor (probability), we fully understand the total process or the roadmap. Having a program manager with appropriate experience should be considered essential for commercialization.

6.2 ENTREPRENEURIAL LEADERSHIP TO BUILDING COMMERCIAL VENTURES

Most often, an entrepreneur has lot of capability in leading the efforts to kick off an initiative and leading the commercializing concept, but they may not have all the knowledge required for developing and running successful commercial ventures. Normally, an entrepreneur is equipped with enough knowledge to kick off the venture, but soon he or she will require other specialists or experts to join in to cover all the other areas to support additional research and development (R&D) capability needs, prototyping, testing, manufacturing, marketing, financing, and managing the venture with adequate profitability for sustainability. Initially, the entrepreneur needs to add people who could help during the R&D phase, but soon he or she would need to have people who could help in getting the financing for the R&D phase, then for raising funds for prototyping and testing, and later for supporting manufacturing and market testing. Entrepreneur's primary job is to provide leadership and create environment that promotes teamwork. Depending on the timing, additional resources/expertise would need to be added and integrated into the team.

The details on the type of expertise to be added and the time of hiring and training should be spelled out in the business plan. The business plan needs to include at least the following sections:

1. **Entrepreneurial culture**: The entrepreneur creates the culture and develops the environment to influence the others involved.
2. **Building an organization**: Entrepreneur's primary function is to provide the shape of the organization and give the direction.
3. **Prioritizing goals and objectives**: Entrepreneur defines the goals and identifies the objectives and prioritizes the goals and objectives.

4. **Effective communication**: Effective mode of communication is the key to building the organization. The organizational team should understand clearly the message, and that should be provided in writing, stated often and practiced regularly.

5. **Inspiration and performance**: The entrepreneurial spirit and concept are the sources of inspiration, and they set the performance standards.

6. **Feedback and rewards**: The organization should be built on receiving feedbacks frequently from the team members as well as from the customers. Accurate and timely feedback should be collected and acted upon quickly. Good feedbacks should be recognized and rewarded.

6.3 TEAMING FOR SUSTAINABILITY

Cash flow and expertise are the key ingredients for any venture, especially for the first-time nanoproduct ventures. These commodities can be brought in by carefully selecting external resources as partners and by offering them certain rewards or consideration of values. This may include sharing a part ownership (thus sharing the risks as well) of the venture and/or sharing profits at a negotiated terms and conditions acceptable to both parties.

1. **Strategy and objective**: The partnering objectives and strategies are created during the business plan development. As the roadmap to nanoproduct is developed and the proposed venture starts taking shape, the partnering strategy starts taking shape. Overall objective(s) may remain the same, but implementation strategies can change.

2. **Needs assessment**: Need assessments should continue at different stages of the roadmap, as the situation(s) do change every so often.

3. **Selection of expertise**: During each phase of nanoproduct development, it is expected to face different challenges. So, the need for different types of expertise will change at different stages of roadmap. The planning process should be flexible enough to recognize and secure expertise as required.

4. **Hire or subcontract**: Most of the time, the needs for additional expertise will be for a short time. So, that will dictate

whether the nanoproduct venture should consider hiring or subcontracting. Consider hiring when the need is continuous and for longer time. Subcontracting is preferable when the need is for a shorter period or as and when required.

5. **Technical collaboration**: When certain activities of a nanoproduct venture are not required on a continuous basis, then it may be prudent practice to consider having technical collaboration with another organization with expertise and adequate resources. Sometimes, collaboration with an organization having market recognition for technical expertise and cost-effectiveness should be considered highly desirable.

6. **Board of directors**: Sometimes, the members on the board of directors are selected because of their expertise. These directors' technical, business, or financial expertise can be great assets, at a minimum cost to the nanoproduct commercialization venture.

7. **Support services (short-term vs long-term)**:
   a. **Individual versus organization**: Quite often, nanoproduct venture struggles with the benefit assessment on having an individual employee or engaging an organization as a subcontractor. The benefits on both options can be difficult to access.

   b. **Accounting and financial**: Most often, it is prudent to engage an outside accounting firm for accounting and financial services.

   c. **Legal and intellectual property**: Also, the nanoproduct venture (or start-up companies) retains contracted services for legal and IP support. This approach can be quite cost-effective.

   d. **Manufacturing**: During the early stage of the venture, the sales are not strong enough to require own manufacturing facility. It may be prudent approach to use external manufacturing facility during early life of the venture. This approach may be costlier than having your own facility. But strategically external manufacturing option is lot more desirable considering financial commitment and expenditures for having your own facility.

   e. **Quality control**: It will be a prudent practice to use external, reputable QA/QC services. A reputable third-party QC report will carry much value with the investors or client community.

f. **Specialty R&D support**: Only on special cases, consideration can be given to using an external specialty R&D support who has reputation in solving a specific R&D issues.

6.4 CUSTOMERCARE ESSENTIALS

Business success is created through the eyes of the customers. Proper customercare help creates healthy revenue and profit performance. Customer experience can influence the perception of the product and viability of the organization. Many organizations have developed a number of techniques on how to improve the customer perception on the product, its pricing, competitiveness, and sustainability. Good strategy is to collect customer feedback, understand the issues, and create the program that will not only maintain the level of confidence of the customers but also improve on it gradually or stepwise. The program must address both tactical and strategic issues.

In the eyes of the customers, pricing of the product is very important. This involves affordability and competitiveness. One of the many challenges is to evaluate the pricing of the competition and keeping lookout for promotional programs of the competition. Based on market intelligence, the timing and content of your promotional programs should be developed and implemented. Also, it is important to understand the impact of your promotional activities.

For any ventures, it is essential to understand why customers buy your product at a price they are willing to pay. So, it is very important to get customers' feedback, so that immediate actions can be taken for any negative comments. It is important to develop a system for engaging the customers regularly. Customers can also tell us the negative aspects of your competitors' products.

So, design the customercare program to collect market intelligence and trending. At the time of introducing the product in the market, the nanoproduct venture needs to develop policies on pricing and promotion. Also, the venture must develop policies and guidelines on customercare and customer feedback.

APPENDIX A1

Nanomaterials Characterization

A1 (1): Graphene

Graphene and graphene-related materials have revolutionized numerous areas of materials science and technology. Their massive technological success is related to their unique structural and chemical properties. Graphene structure and its key fundamental properties such as surface area, pore size, and density are examined.

Graphene has the largest surface area to volume ration among novel 2D crystalline-layered materials. Being spread as an atom-thick layer of hexagonal ring-bound carbon atoms, all atoms in a graphene sheet are surface exposed. This gives graphene a rich array of unique surface physical, chemical, and electronic properties that continue to open doors for new applications in nanotechnology and energy sectors.

Surface area: Surface area impacts every application of graphene and graphene-related materials (such as graphene oxide, graphene–metal oxide composites, heteroatom-doped graphene, and nanostructured photocatalysts, among others). It is largely the exposed surface of these materials that interacts with gases, liquids, solids, electrons, ions, and photons. Therefore, evaluation of the surface area of graphene is a crucial step in understanding and optimizing their performance.

Graphene sheets, if fully exposed and reasonably large, have a theoretical surface area of $2{,}629\,\mathrm{m^2/g}$. Surface areas of that magnitude have indeed been reported following activation of exfoliated graphene oxide. However, graphene sheets tend to stack one on top of another due to weak, but extensive interactions between their surfaces. Graphene layer stacking reduces their accessible surface area in proportion to their degree of stacking.

Pore size: Pores in graphene or graphene-related materials can include holes within sheets, whose dimensions can be tailored by things such as selective ring removal and nitrogen passivation. It can also include spaces between sheets, with the overall pore dimensions and size distribution being dictated by the degree of stacking, crumpling, or pillaring with additives.

Density: Gas pycnometer provides a fast, clean, and nondestructive way to assess the density of carbon materials in general. The precision and accuracy of modern gas pycnometer are adequate to assess the differences in chemical and physical characteristics of graphene-related materials. The density of graphene sheets can increase with increasing stacking order and perfection. Perfectly stacked and aligned graphene sheets have a density close to that of crystalline graphite. However, heteroatoms, stacking imperfections, and defects tend to lower the density to a value that depends on heteroatom nature and content and on pore characteristics. In some cases, pores created during stacking or agglomeration can remain closed.

Reactivity: Although the ideal graphene 2D crystals have uniform surface, real graphene materials are often energetically, chemically, and physically heterogeneous. Surface sites that may be more reactive toward adsorption, ion or electron exchange, and mechanical strain include graphene sheets edges, heteroatoms, functional groups, impurities, and metal catalysts.

Summary: Graphene and graphene-related materials are currently at the forefront of materials research and technology development. The precise evaluation of their structural characteristics is an essential step toward optimizing their performance. Specific properties that affect virtually every application of graphene materials include their specific surface area, pore size distribution, density, and reactivity. Graphene reactivity is related to the nature and concentration of active sites, which can be quantified using chemisorption and temperature programming techniques.

(Note: The aforementioned information has been adopted from Anton Paar GmbH publication.)

A1 (2): Microporous Materials

Microporous materials are defined as materials containing pores that have diameter less than 2 nm. The surface area of porous materials is commonly calculated utilizing the Brunauer–Emmett–Teller (BET) equation. The BET equation is applicable for surface area analysis of nonporous, macroporous, and mesoporous materials and is applied in the relative pressure range P/Po from 0.05 to 0.30.

The BET method is based on the assumption that a monolayer of absorbate is formed on the pore walls. In materials possessing micropores, it is difficult to differentiate between the monolayer formation process and the micropore filling process that usually occurs at P/Po below 0.15. Sometimes, BET method may lead to a significant overestimation of the monolayer capacity. For these reasons, the surface area obtained using the BET equation for microporous materials should be considered a "characteristics" or "equivalent" surface area, or it should just be called BET area. The BET method in a strict sense is not applicable in the case of microporous adsorbents.

It is important to note that while the method allows one to very quickly determine the linear BET range for a microporous sample, surface areas obtained with this procedure are still considered apparent or equivalent surface areas. This method considerably improves the comparability and reproducibility of results between different labs and within the literature.

(Note: The aforementioned information has been adopted from Anton Paar GmbH publication.)

APPENDIX A2
Guidelines to Selecting Manufacturing Process(s) for a Nanoproduct

The process involves reviewing the potential options. It begins with assessment of available process technologies to choose from, identifying any innovation requirements, selecting a process, and then integrating the process into the total system. The assessment of needs requires having industry standards and guidelines; understanding of technology readiness, if new technology is desired; having past experience in commercializing technology products; prototyping/additional testing requirements, if appropriate; assessing appropriate risks for the selected process; and then preparing a strategy for risks mitigation.

Manufacturing process selection methodology includes the following:
1. Clearly identifying APPLICATION
2. Defining NEEDS and Selecting TECHNOLOGY(S)
3. Selecting Manufacturing PROCESS(S) System(s)
4. Describing COMPETITIVE-ADVANTAGES
5. Assessing Appropriate Technology's READINESS-LEVEL
6. Resolving Commercialization CHALLENGE(S), if any
7. Identifying RISKS and Developing Risk MITIGATION Strategy

8. Completing Technology LICENSING Arrangement, if appropriate
9. Developing Manufacturing System DESIGN and Training of Operating TEAM
10. Building actual manufacturing SYSTEM

The manufacturing processes that have been used in the laboratories for nanoproduct commercialization include at least the following processes. Sometime, a combination of processes would be necessary to achieve the ultimate result.

1. Biological
2. Combustion
3. Deionization
4. Electrospraying
5. Flow synthesis
6. IR spectrometry
7. Laser deposition
8. Microwaving
9. Morphology manipulation
10. Nanofiltration
11. Separation
12. Classification
13. Synthesis
14. Pyrolysis
15. Vortex mixing

Once the preferred option(s) are identified, describe the competitive advantages of each process option. Ultimately, this would involve optimizing the design and operation. The target objectives would include the following:

1. Increased reliability and profitability
2. Created competent management team
3. Secured sustained competitiveness
4. Achieved resource conservation
5. Obtained lowest overall costs of operation
6. Secured minimized pollution and maximized wastes utilization

7. Achieved creating jobs and local industrialization, if possible
8. Led socioeconomic development in local market and region, where possible

Once the "best option" is identified or selected, then its advantages are to stated and written down, its limitations/potential difficulties are to be recognized, and then risks are identified and risk-mitigation options are recognized.

BIBLIOGRAPHY

Altmann, Jurgen *"Military Nanotechnology"*, Routledge.

Anwar, Sohail, Raja M., Yasin Akhtar, Qazi, Salahuddin *"Nanotechnology for Telecommunications"*, CRC Press.

Auplat, Claire *"Nanotechnology and Sustainable Development"*, Routledge.

Brechignac, Catherine (Editor), Houdy, Philippe (Editor), Lehmani, Marcel (Editor) *"Nanomaterials and Nanochemistry"*, Springer.

Brozek, Tomasz (Editor) *"Micro- and Nanoelectronics: Emerging Device Challenges and Solutions"*, CRC Press.

Caves, Richard *"American Industry: Structure, Conduct, Performance"*, Prentice-Hall Inc.

Challa S. S. R. Kumar, *"Nanomaterials for Biosensors"*, Wiley-VCH.

Connellan, Thomas K. *"How to Improve Human Performance – Behaviorism in Business and Industry"*, Harper and Row, Publisher.

Crawley, Tom *"Commercialization of Nanotechnology – Key Challenges"*, Nanoforum.

Critchley, Liam *Nano Magazine*, e-Newsletter.

Ganguli, Prabuddha, Jabade, Siddharth *"Nanotechnology Intellectual Property Rights: Research, Design, and Commercialization (Perspective in Nanotechnology)"*, CRC Press.

Haack, Stefan, Burghardt, Irene, *"Ultrafast Dynamics at the Nanoscale: Biomolecules and Supramolecular Assemblies"*, CRC Press.

Helwegen, Wim (Editor), Escoffier, Luca *"Nanotechnology Commercialization for Managers and Scientists"*, Pan Stanford Publishing.

Hicks, Herbert G. *"The Management of Organizations: A Systems and Human Resources Approach"*, McGraw-Hill Book Company.

Holloway, Charles A. *"Decision Making under Uncertainty – Models and Choices"*, Prentice-Hall International Inc.

Hosono, Hideo (Editor), Mishima, Yoshinao (Editor), Takezoe, Hideo (Editor), Mackenzie, Kenneth *"Nanomaterials: Research towards Applications"*, Elsevier Science.

Jackson, Mark (Editor), Morrell, Johnathan (Editor), *"Machining with Nanomaterials"*, Springer.

Jain, Vijay Kumar *"Nanofinishing Science and Technology: Basic and Advanced Finishing and Polishing Processes"*, CRC Press.

Ke, Changhong *"Recent Advances in Nanotechnology"*, Academic Press.

Kimmons, Robert L. *"Project Management Basics"*, Marcel Dekker, Inc.

Kimmons, Robert L. (Editor), Loweree, James H. (Co-Editor) *"Project Management – A Reference for Professional"*, Marcel Dekker, Inc.

Krishnamoorthy, Sivashankar (Editor) *"Nanomatrials: A Guide to Fabrication and Applications (Devices, Circuits, and Systems)"*, CRC Press.

Lau, Woei Jye, Ismail, Ahmad Fauzi, *"Nanofiltration Membranes: Synthesis, Characterization, and Applications"*, CRC Press.

Manafield, Elisabeth (Editor), Kaiser, Debra L. (Editor), Daisuke, Fujita (Editor), Marcel, Van de Voorde (Editor) *"Metrology and Standardization of Nanotechnology: Protocols and Industrial Innovations"*, Wiley-VCH Verlag GmbH & Co. KGaA.

Mathur, Rakesh Behari, Singh, Bhanu Pratap, Pande, Shailaja *"Carbon Nanomaterials: Synthesis, Structure, Properties"*, CRC Press.

Mensah, Thomas (Editor), Wang, Ben (Editor), Bothun, Geoffrey (Editor), Winter, Jessica (Editor), Davis, Virginia (Editor), *"Nanotechnology Commercialization: Manufacturing Processes and Products"*, Wiley-AIChE.

Musa, Sarhan M. *"Computational Finite Element Methods in Nanotechnology"*, CRC Press.

Rickerby, David (Editor) *"Nanotechnology for Sustainable Manufacturing"*, CRC Press/Taylor & Francis Group.

Roco, Mihail, Mirkin, Chad A., Hersam, Mark C. *"Nanotechnology Research Directions for Social Needs in 2020: Perspective and Outlook (Science Policy Reports)"*, Springer.

Schodek, Daniel L., Ferreira, Paulo, Ashby, Michael, *"Nanomaterials, Nanotechnologies and Design: An Introduction for Engineers and Architects"*, Amazon.

Shah, Mohammad Ashraf, Bhat, M. Amin, Davim, J. Paulo, *"Nanotechnology Applications for Improvements in Energy Efficiency and Environmental"*, IGI Global.

Shatkin, Jo Ann *"Nanotechnology"*, CRC Press.

Shimazu, Shogo, and Tursiloadi, Silvester *"Transferring Nanotechnology Concept towards Business Perspectives"*, Daya Publishing House.

Slingerland, Janet *"Nanotechnology (Cutting-Edge Science and Technology)"*, Essential Library.

Sparks, Sherron *"Nanotechnology"*, CRC Press.

Tibbals, Harry F. *"Medical Nanotechnology and Nanomedicine"*, CRC Press.

Tsakalakos, Loucas *"Nanotechnology for Photovoltaics"*, CRC Press.

Tsuzuki, Takuya *"Nanotechnology Commercialization"*, CRC Press.

Victor, Sontea *"3rd International Conference on Nanotechnologies and Biomedical Engineering"*, Springer-Verlag Hmbh.

Vo-Dinh, Tuan *"Nanotechnology in Biology and Medicine"*, CRC Press.

Waitz, Anthony, Bokhari, Wasiq, *"Nanotechnology Commercialization Best Practices"*, Quantum Insight.

Werbach, Adam *"Strategy for Sustainability – A Business Manifesto"*, Harvard Business Press.

Wilson, Mick, Kannangara, Kamali, Smith, Geoff, Simmons, Michelle, Raguse, Burkhard, *"Nanotechnology"*, Chapman and Hall/CRC.

Zhang, Wei-Hong, Maguire, Russell, Dang, Vivian T., Shatkin, Jo Anne, Gross, Gwen M., Richey, Michael C. *"Nanoscience and Nanomaterials: Synthesis, Manufacturing and Industry"*, Amazon/DEStech Publications.

Printed in the United States
by Baker & Taylor Publisher Services